# *Really? Wow!*

**An extraordinary projection of what will become of the human race.**

**Bill Thurston**

# Really? Wow!



For information about the contents of this book, or permission for reproducing portions of this book, please email the author at **thurstongroup@cox.net.**

Published in the United States

ISBN-13: 978-0-9886544-3-3

## Table of Contents

Introduction 1
1. What is the process we will use? 4
2. Fields of Interest 9
3. Human Physiology Past to Present 12
4. The Computer Past to Present 18
5. Robotics Past to Present 23
6. Wireless Communication Past to Present 26
7. Mathematics & Physics Past to Present 28
8. Human Perception Past to Present 31
9. Connections Between These Fields Today 42
10. Why Use a Period of 20 Years? 45
11. Human Physiology in 2032 46
12. The Computer in 2032 61
13. Robotics in 2032 64
14. Wireless Communications in 2032 66
15. Mathematics and Physics in 2032 69
16. Human Perception in 2032 71
17. Connections Between These Fields in 2032 75
18. How will these changes affect us in 2032? 79
19. Human Physiology in 2062 83
20. The Computer in 2062 85
21. Robotics in 2062 91
22. Wireless Communication in 2062 95
23. Mathematics & Physics in 2062 96
24. Human Perception in 2062 98
25. Connections between these fields in 2062 100
26. The Big Dilemma 102

Acknowledgments

Thanks to my wife, Debbie, who allowed me to go to my "cave" for over a year to create this book.

Thanks also to family, friends, and business acquaintances who provided much of my knowledge that enabled this book to happen.

# Introduction

Maybe the most positive, exciting, new, major event in humankind history will be upon us in less than five decades. It may truly be a start of a new phase for human existence. I believe all of us, as a world society, will willingly embrace the discovery and make basic human changes that will support a much better world for us and future generations.

We have all the information we need today to be able to present this discovery with a very simple projection process. More detail about this projection process will be presented in the next chapter.

Why use the word "projection" instead of "prediction?" Well, a prediction is a statement about the way things will happen in the future, often—but not always—based on experience or knowledge. So anybody could make a prediction about anything! A projection is a prediction based on known evidence and observation. That is the process used in this book.

You might ask, "Doesn't this require thousands of scientists working for decades in many fields in order to come up with this type of major human discovery?" Of course the answer is "yes" and we do have countless people around the world improving human understanding on a daily basis. The key is that nobody has evaluated the convergence of many complicated sciences to make all of us aware of this discovery that is almost upon us. I guarantee you that if somebody had published this new discovery, you would have already read about it or heard about it through the media.

We will address six very complex fields of interest in such a way that all of us can understand it independent of any person's particular technical understanding of the six subjects. Introduction to these six fields will be presented in Chapter 2 titled "Fields of Interest."

Before getting started, there are a set of simple intellectual positions that are used to produce everything in the book. Here is a list of those positions.

1. All thoughts in this book stem from our human senses and the brain power we possess to understand our environment.

2. In order to have a future understanding, you need to start with having some present understanding.

3. A human's ability to reason is a key to the understanding presented in this book.

4. A human's ability to develop tools is a key to the understanding presented in this book.

5. Understanding the history of human discovery is a key to unlock next steps in discovery.

6. Change in human physiology was important to us in the early years of our existence. Things like developing an opposable thumb, walking upright, an increase in brain size, all contributed to our position in the world today. Growth in human understanding today depends very little

on humans' changing physiology. Today human understanding comes from our mental abilities to create tools and to develop new ideas.

7. The growth of human understanding today follows a simple pattern.

Here is that Human Understanding Process:

a. We investigate things or by chance something new is brought to our attention.

b. We then discover something and can rediscover it over and over again.

c. We experiment with it and learn more about it.

d. We understand it and define it. We do this with our mathematics and physics.

e. We use it to our advantage which triggers more investigation.

f. We make tools based on the discovery.

g. The discovery becomes part of human understanding which allows new investigations and discoveries.

While reading this book, you may not believe or not want to believe one of the assumptions or projections. At that time you could put the book down and not finish the journey. I hope your curiosity allows you to finish the journey; but if not, maybe in the future as humans roll out new discoveries, you may want to take a second look.

Many of today's events in human society are very troubling and we are eager to find solutions. I believe solutions to many of our societies' problems will be addressed and changed for the overall health of human society with the new discoveries presented in this book.

Chapter 1

## What is the process we will use in order to make our projections?

We will start with six areas of interest that will reveal the new discoveries:

- Human Physiology
- The Computer
- Robotics
- Wireless Communication
- Mathematics and Physics
- Human Perception

Okay, I can hear many of you thinking, "I don't know a lot about all of that." Don't worry, you will get it.

We will use a projective process because that is what we are doing—making a projection of a future event. We use projective processes every day of our life. You start with a present understanding and project a future event that will happen.

Here are examples:

- If I continue to do well in high school, I will be eligible to go to college.

- If I take one more golf lesson and practice at the driving range every Monday and Friday, I will be able to break 90 by October.
- If I join Zumba next week and go to all classes, I will be able to fit into all my bathing suit this summer.
- If I get home on time, I will be able to watch Monday Night Football.

Now sometimes this process turns into a two- step process. First, I will understand the present situation. I will project a future event that is required before you make the final projection. That is the process we will use in this book. It is hard to find examples of this because individuals rarely use this process in everyday life.

Here are two examples:

- I want some ice cream. Step 1: Go to the ice cream shop. Step 2: I decide which flavor I want and buy it.
- Another good example will be easily understood by people who have played adventure video games. You constantly investigate, discover, adapt, rediscover, and grow your character and the characters' successes.

Okay, that's the idea. Now here is the exact process.

The approach in creating possible future understanding is to use humans' present tools and understandings and extend and combine that information with what logically might be understood in the future. Resulting understanding should be verified and then be used to create new tools.

The Projective Process we will use is a simple multi-step process.

1. We analyze the six areas of interest and present our present level of understanding of each and how they interact today.
2. We will project the state of each area of interest into the year 2032.
3. We then view the world as it will be in 2032. How would the set of projections have changed our lives and our understanding of the world in 2032?
4. We take our world as it would be in 2032 and make another set of projections 30 years in the future to the year 2062. These projections most likely will have a lesser degree of certainty than the first projections.
5. We then view the world as it will be in 2062. How would the set of projections have changed our lives and our understanding of the world around us.
6. We then draw conclusions about our understanding 50 years in the future.

That is the complete process.

Here is an example of using this projective process using just one area of interest.

Let's choose the area of interest to be the amount of memory in personal computers. Over the last ten years, we have watched memory in personal computers constantly increase. The first computers provided you information only in the form of text. There were many reasons for this but limited memory size was one of the limiting factors. The first disc drives were not capable of storing even one high definition picture. As the size of memory increased, computers were capable of displaying pictures and today, with present memory, computers commonly use text, pictures, animation and videos.

Even if you don't understand much about computers, you could project with some certainty that the amount of memory will increase in new computers in the near future. If you had a high degree of understanding of ongoing research about computer memory technology, you would have a high degree of confidence that new computers will have more memory. If you were a computer engineer working in a company that makes memory, you may already have seen prototypes of new, larger memory working in computers today. In this case you would have a very high degree of confidence that the new technology would be in commercially-available products soon.

With this simple example, it is quite easy to project that computers will have more memory in ten years with a high degree of certainty and without assuming too much.

Next, we imagine ourselves in 2032. What new things might we be doing with this larger memory in our personal computer?

Here is an example of a conclusion after the first projection. Let's say the average home computer would have ten times more memory.

Imagine how everyday life would be affected by this advancement in memory technology.

A relative living on the other side of the country has a baby and wants to give you information about this great event using computer communications.

- With only text (1980s): Baby girl. 14 pounds. 22 inches long. Curly hair.
- With pictures (1990s): Picture of Mom and Dad with the new baby. Several close up pictures of the baby.
- With video (2000s): You experience the baby moving and maybe crying along with voices of the proud parents.
- Near term projection (2020s): 3D real time images without 3D glasses. Better voice communication. Full 3D audio video notes.

In the 1980s, pictures on computers would have been a first-level projection and videos on computer would have been a second level projection.

In the 1990s, videos would have been a first level projection and 3D images on your computer without glasses would have been a second level projection.

Today 3D images without glasses are a first level projection because Sharp Laboratories showed a prototype at the Royal Society in London in 2010. At this time, I don't know of a second level projection from today.

That's all there is to the process. Pretty simple, but the work is in the projections.

This book was written using this process.

Chapter 2

# Fields of Interest

**Human Physiology** is the first field of interest. We will focus on how our body works. In particular, we will focus on how the brain works and how it works with the rest of our body. Also, we will focus on genetics, in particular, the understanding of our DNA which is the place where all genetic information of an organism is stored. We will address the connections between humans and the machines humans build.

**The Computer** is the second field of interest. I think a short history of its development is important. Showing how its operation is tied to human perception will be presented. We will discuss how we communicate with computers, and give a high level view of how computers work. We will address changes to be made to computers to emulate our growing understanding of how the brain works.

**Wireless communication** is the third field of interest. This technology has firmly entrenched itself in our lives: cell phones, car radios, GPS, the internet, radar, communications satellites, and much more. For example, the internet may be enhanced in the future to communicate similarly to how the brain communicates. We developed and have been improving the internet for about 50 years but the brain has been around and improving itself for millions of years. Certainly we have much to learn in this field of interest.

**Robotics** is the fourth field of interest. What might first come to mind are mechanical machines that try to act and look like humans. Most robots today don't look anything like us but play a large behind-the-scenes role. Robots make about everything in large manufacturing companies. Aircraft drones are robots that are presently replacing many of our manned fighter aircraft. Even something like a car or bike is a form of robot controlled by the human brain. We will explore the growing relationship and dependency on robots.

**Mathematics and physics** are the fifth field. We will look at these tools today and explore how and why they will change in the future.

**Human Perception** is the sixth field of interest. Here we will talk about local reality, universal reality, and how we think and synthesize information. Today most of our perception is from the location of our physical body. Of course, TV and movies give us a perception from a position defined by the producer. In the future, your perception could be moving to points physically different from where your physical body exists.

In the process of looking into these areas, we will touch on some other areas of interest like addressing life beyond Earth and where to look for new signs of new local and universal order. Also we will see if there is a universal reality in which we can participate.

All of this will tie together to make the final projections in the last chapters.

Okay, let's review each field of interest as it exists today along with a little history of each.

Chapter 3

# Human Physiology Past to Present

We have learned quite a lot about ourselves. We know how to cure some diseases. We know how to set fractures, replace blood vessels, treat some infections, perform heart transplants. We know about genetics and how traits are passed from generation to generation. We know the basic code that generates us is found in our DNA. We know all living things on Earth have similar DNA. We know how to splice genes and clone animals and plants.

Today many people don't like the idea of humans splicing genes on humans to make “man-made people” but much work is going on around the world. The field of understanding DNA is at such a level that we could stop the passing of hereditary diseases from generation to generation. We actively modify plants for many reasons, including making harvesting easier, so they are resistant to insects or diseases, so they grow bigger or smaller, and much more. Almost all hard cheese is cultured with a genetically-altered rennet. We are actively modifying animals for our own reasons. We are creating new breeds of animals that are in some way better for manufacturing human food. We are doing more bizarre things by combining parts of one animal with another. For example engineers have equipped *E.coli* bacteria with the spider genes responsible for spider silk production. The result is synthetic spider silk produced by the bacteria. The silk could be used for fine sutures, replacing tendons, and tough yet light textiles. Whether they are called biotechnology companies, genetic companies, or genetic research companies, they are very active in creating life modified by us. The top three U.S. companies in this field are Amgen, Genentech, and Genzyme and they employ over 30,000 employees.

Many of us have a good example of humans modifying animals for their own reasons right in their own home. For those that have a dog or cat, look at that animal. Would nature have ever created such a domesticated animal without human intervention?

We also know about our human senses. They are our senses because the human brain recognizes them uniquely. In computer lingo, our senses are inputs to our brain from different "input devices" or organs. Eyes for "seeing." We know we only see a limited light spectrum. Other animals like pit vipers can see heat. Our ears hear a limited audio spectrum. We know we are limited compared to other animals in hearing range, hearing intensity, and understanding the location of the sound source. Our nose for smelling. We have limited ability to smell as compared to other animals. Sense of feeling. Sense of knowing where our body parts are in space. Sense of pain or pleasure. Balance and acceleration. Temperature. We know of senses used by other animals like echo location, generating or sensing electric or magnetic fields. We generally know about each sensing organ and that they send information to the brain on human wires called nerves.

We know the human brain accepts a continual stream of input and responds to some of it, stores some of it, and rejects or ignores most of it.

Stored information is synthesized uniquely by each person. For example: Our eyes "see" a very complex set of information. Our brain selects information as being important or not important. Important information may be saved, but certainly most of it is rejected. Imagine you got in a car and drove 100 miles, storing every piece of information you saw—every white line, every cloud, every car with its model, color and license plate number, every street name, every plant and animal. We know we don't do this.

Each person has their own filter for remembering information they sense based on the "importance" to them. An example of this would be to blindfold several people and take them to an unknown spot and then remove the blindfold for ten seconds, then blindfold them again and ask each to recall the five most important things they saw. Each will likely recall differently.

How we store information in our brain is not fully known to us today. We do know we can memorize or force storage of information. We know we can focus at different levels. An example would be to take a test sober then take it inebriated or sleepy. We know we can recall things independently of our conscious effort. We all have had something just "pop" into our mind. In other words, our conscious state doesn't control all systems within us. Could you imagine if we consciously had to control our heartbeat!

The human brain associates time with the information it stores. Some of the time information is exact time as we define it and some is relative time. Examples of exact time would be remembering a birthday or what time your favorite TV program is aired. Examples of relative time are knowing you met Mark before you met Jim or you went to the store right after taking a shower.

Our primary area of interest in this huge field of physiology is the human brain and how it works. We know quite a lot about what it does but much less about how it works.

Here is a short list of what we do know and will use in our projections:

- We know we have a conscious state.
- We know we dream (a different state from the conscious state).
- We know there are subconscious activities that control some body and mind functions.
- Humans can do many things at the same time. Processes happen in parallel and seem simultaneous.
- Stored information in the brain has been synthesized based on who you are.
- Stored information is recalled with associated information.

- Our brain is selective in storing or in doing any advanced processing on sensory data.
- Our process to find importance in data from senses is a separate process.
- We know we can respond to a stimulus (like a hot stove) and quickly move with what seems like little conscious intervention.
- We can memorize. This is a conscious effort to store information. Consciously stored information seems to have very little associated information. An example: What is 8 times 8? Not much "popped up" except 64. How was your vacation? Lots of information and secondary information "pops up."
- We seem to operate at different levels of intensity.
- Automatic responses vary for each person. Example: Try scaring somebody. Some will scream, some will prepare to defend, some will attack, some will retreat.
- We aren't consciously in control of all our processing (standing, breathing, dreaming).
- Humans take in everything through senses and use only a very small part of the information.
- Human emotions. Human emotions are always "on." You can't turn them off. You simply create the memories that create the emotions.
- Human profiling is always on. As we are constantly sensing our environment, we are also constantly comparing the new set of senses to our memories to take appropriate mental or physical actions.

These are the areas of human physiology we will investigate for our projections.

Chapter 4

## The Computer Past to Present

We don't need to spend a lot of time talking about computers. Most of us use them for various reasons. They are in our cars, phones, laptops, desktops, mobile devices, televisions, printers, cable and satellite boxes, and many other places.

Computers basically listen to input devices like keyboards, mice, scanners, cameras or other types of instrumentation and then based on some application, output information to monitors, speakers, printers that is intended to be useful to the user.

Computers started out in the 1970s with tiny internal memories of maybe 8,000 bytes and speeds of less than one million instructions per second with maybe five megabytes of disc space.

Today, only 40 years later, a home computer can easily have 16 gigabytes of internal memory, speeds of 4,000 million instructions per second with maybe two terabytes of disc space.

These massive increases in speed (one thousand to one million times bigger and faster) have allowed us to now have pictures, videos, realistic animation, and artificial voice creation as common computer experiences. If you have ever experienced the Xbox Kinect games, it is quite amazing. It responds to your slightest movement to do things like teach you intricate dance moves by evaluating your movements to music.

Maybe the computer is one of our first tools to bring a new type of understanding to humans because humans can make computers do things that humans can't do and will probably never do. A good example is the distribution of fleets of trucks (maybe 1,000 trucks) to accomplish their goals of moving freight in the most efficient way. No human or group of humans could figure this out in a timely enough fashion.

Another example is an electronic eye connected to a computer examining potato chips moving by on a conveyor belt. Moving by at the rate of 1,000 chips per second, the chips need to be evaluated in order to meet some specification of color, size, and maybe shape. Once the computer "sees" a bad chip it directs a blast of air to remove it from the good chips. All this happens without the conveyor belt ever stopping. No human or set of humans could do this.

Today's computers are called "digital" computers. In the very deepest bowels of every digital computer, all you have is millions and millions of "on/off" switches that can be turned on or off in less than a billionth of a second. Combining this sophisticated tool with programs that were created by the human brain gives humanity arguably the most powerful human tool ever.

Do humans make computers in their own image?

All computers are programmed to do what they do based on a program created by humans. This includes how we think and how we synthesize data. Flow charts of computer programs are very understandable and precise.

As amazing as the human brain appears to us, it is anything but precise. We forget things, we all view the world differently, our mind/body connection is different. Some people are athletes and some are "klutzes."

Maybe the human brain works in a way that couldn't be defined with traditional flowcharting. Maybe there is additional knowledge to be unlocked in our environment by computers but we don't find it because we don't yet have the knowledge to program the computer to "look" for it.

When you think about it, man operates the exact opposite of today's computers. Man takes in everything through senses and only uses a very little part of it. Today, our computers take in a relatively small amount of information and try to explain something that is universally true.

Another major difference between the human brain and a computer is the amount of energy required to make them work. The human brain is considered to be a sophisticated "computer" doing certain operations. To duplicate the ability of the human brain would require a computer that uses much more energy than a human body could support. Therefore the human brain must be significantly more efficient than our existing computers.

The magnitude of this difference suggests that at the core of each, they are very different.

Here is a list of present human understanding of computers:

1. Humans have created this logical electronic device we call a computer.
2. It is massively faster than the human brain doing mathematical operations. Maybe there is a problem that humans would like to solve and can define but know it would take a million engineers a million years to solve. A present-day computer could likely solve the problem in less than one hour.
3. It can monitor and record events with computerized instruments without humans being present.
4. It can operate in places where humans cannot. Corrosive environments, small places, large pressures, places with no human food, etc. The Mars Rover computers operates just fine on a Mars surface hostile to humans.
5. It can share and condense information for humans.
6. It can recall past information much faster and with more accuracy than humans. We take some of this capability for granted. Let's say you want to know about the American bald eagle. If you go to your favorite search engine, you can get a massive amount of information about the American bald eagle in a few seconds. How would you do it without the computer?

7. You can think of the computer as the tool that most represents the sum of all scientific human understanding. It will continue to support this role well into the future.

Computers will be a key contributor to the projections that will be made in this book.

Chapter 5

## Robotics Past to Present

What are robots? They are manmade devices directed to do something by a human or computer. I think most of us have an image of a mechanical human-like thing when they think of robots or the robots that help us make products like cars or appliances. But is a car a robot? You could always walk to the store, but if you use a machine like a bicycle, motorcycle, or car you are actually using a robot. You provide the brains to direct the vehicle but it does the rest.

Robots can be very advanced today and here is just one example. There is a new type of robot called a Lingodroid that is programmed to share language. The robots can coin phrases to describe places they have been, places they want to go, and plans for getting there. When they need a new word, they invent one. For example, in one game, two Lingodroids roamed through a course and met in an unfamiliar part of the course. The meeting triggered one robot to name the place “jaya” and shared the new word with its partner who then added the word to its lexicon. In this way the robots slowly build a new language to describe their travels. Doesn’t that sound similar to human behavior?

Robots can do very specific predefined activities very well and in many cases much better than humans.

Factories use robots to perform tasks such as welding, assembly, sealing and operating dangerous tools. They never tire so they can perform their jobs 24 hour a day.

Robots are used in military applications. As unmanned aerial reconnaissance vehicles, robots can enter zones of operation that traditional aircraft cannot. Robots have been used for situations in which no human could survive. The *Pioneer* robot was able to explore the site of the Chernobyl disaster to assess structural stability and radiation levels that otherwise would have killed a person.

The *Dante II* has been used to enter active volcanoes to research sulfur and gas levels.

Robots are used in space. The Mars rovers *Spirit* and *Opportunity* have been able to continue their missions on Mars for a much longer period of time than would have been possible for a manned mission.

The *Deep Impact* probe that crashed into a comet would never have been feasible for a manned flight to complete, as the robot was on a one-way mission.

Robots are used in medicine. They perform delicate surgical operations. They are used for repetitious time-consuming work done in medical laboratories.

Robots can be used for human convenience. One example is a vacuum robot that cleans the floor of your home. Some household appliances are robots, including toasters, microwaves, and stoves which all support microprocessors that function as safe and efficient devices.

Robots have been developed into incredible human tools, yet there is a whole set of human functions that robots can't perform today. A simple example is to have a robot climb a tree. There is no robot that can do this today. Just think about the complex thought we easily execute when climbing a tree thanks to the human brain.

There will be a major change in robots which will be discussed later in this book.

Chapter 6

## Wireless Communication Past to Present

At first, you might think wireless communication is a new thing that is barely 100 years old, but it has been with us for thousands of years. How about smoke signals, flashing mirrored surfaces, semaphore flags, and lanterns being placed on a watchtower. These are all example of wireless communication.

But if you are talking about humans creating radio waves as a form of communications, it has been around for only 100 years.

Today, we have radio, TV, telephones of all types, radar communication, satellite communication, GPS communication and more.

Now, billions of humans are connected wirelessly with each other and communicate daily.

We might perceive wireless communication as being a much simpler technology than, say, computer technology, but that would not be true.

The speed, accuracy, range, and volume of wireless communication traffic we have created in less than 100 years is truly impressive. We can transmit all the world's dictionaries from one place to another in a few seconds with no errors over a distance of millions of miles. This is a truly amazing human accomplishment and it is available for anybody possessing the right tool, anyplace in the world.

This technology will be much more interesting to us in 20 years as we will discuss later.

Chapter 7

# Mathematics & Physics Past to Present

Some of you may cringe when you hear the words mathematics and physics so this will be brief.

Many, many years ago as humans began using their surroundings and discovering and building tools, there was a need to document their discoveries for future generations.

The documentation of our physical discoveries is what we call "physics" and physics is explained using a tool called "mathematics." That's about it! There is nothing sacred, concrete, or universal about mathematics or physics.

Many past beliefs about mathematics and physics have been totally wrong and changed to be more correct to support new human understanding. There is no reason to believe it is correct today and will surely change to reflect new understandings of the near future. The prediction presented by this book may require a new addition to our set of mathematics and physics tools.

**Do our mathematics and physics explain any universal "laws"?**

We have a weak understanding of universal laws. We have only recently begun trying to understand very little things and very big things. Very little things are the building blocks for the world we experience. Very big things are solar systems, galaxies, and whatever is way farther beyond. Very little and very big things cannot be experienced by human senses so we need tools or machines to help us understand. For very big things, we have telescopes of various types. For very little things, we have microscopes of various types. We have other very large and powerful machines to smash small things together and evaluate what happens in order to gain understanding.

We know of something we call gravity and it appears to be universal in nature, but nobody today can really explain how it works and what it really is. We do see the results that lead us to believe in its existence, but it's still to be understood.

Without getting too serious here, our primary issue with really big or really small things is how we have defined and used time to document human understandings and inventions. Certainly our mathematics and physics are good enough for our normal day-to-day activity, but they are defined using our present concept of time. Why can't certain things go faster than the speed of light? How do we get from the idea of one thing to the idea of two things without a clear mathematical definition of time?

In conclusion, the best way to look at mathematics and physics is that they are a human-made tools to explain what we experience so we can make a more advanced tool in the future. As we learn and discover more, our mathematics and physics will be revised to document these discoveries and this process will continue as long as we pursue new discoveries.

Chapter 8

# Human Perception Past to Present

This section on perception is a little long and sometimes even boring, but is important because perceptions play a large role in coming to the conclusions reached in this book. So, here we go.

Perceptions can be divided into two types. The first is local perception, which we will call “local” understanding. The second type is universal perception, which we will call “universal” understanding. Local understanding applies in our everyday lives. The local tools we use are based on this understanding. Universal understanding applies anywhere in the universe. Some of our local tools and local understanding may also be universal.

Universal understandings must pass the test of being true throughout the universe no matter how big or small. Our present mathematics and physics do well in explaining things we perceive today in our lives, but that is because they were developed based on what we perceive. As we investigate very large things like our universe, which likely hold universal understandings, our tools seem incomplete. Even things we experience and believe to be universal like gravity and magnetism, are not yet fully explained with our present tools.

The goal of looking at our perceptions is to find new tools to extend our ability to learn more about our universe.

In order to discuss perceptions, there are three main pieces to address.

1. We need to know how human sensory organs are used to experience or perceive our world. Also how the human body organs, like muscles, work that move us around. You may think of this as our body's input and output devices. These are things like eyes, ears, nose, those organs that provide chemicals the body needs, vocal cords for voice, muscles, etc.

2. We need to know how these human organs translate the perceived world to "signals" for processing, mainly by the brain.

3. We need to know how the brain processes information, how that information gets stored, and how we make decisions based on that information.

What tools are we using to create our perceptions today? There is no reason to try to define the complete set of all tools used to create perceptions at this time. New tools can be added as discovered.

The first set of tools is that which seems to be available because of the physical aspects of our body. Let's call this tool set our **Physical Tools**. Our five senses provide sensory input when they react to stimuli and transmit signals to our Central Nervous System. These are classified into five types of sensory receptors:

1. Mechanoreceptors
2. Photoreceptors,
3. Chemoreceptors
4. Thermoreceptors
5. Electroreceptors.

Mechanoreceptors detect and respond to hearing, balance and stretching. Photoreceptors are, in a word, our eyes. Chemoreceptors are associated with smell, taste, and the digestive and circulatory systems. Thermoreceptors detect and respond to heat and cold. Electroreceptors detect and respond to electricity. Common terms for these tools are hearing, sight, smell, taste, and touch.

The second set of tools is the perceptions that stem from our ability to remember previous perceptions created by our physical tool set and how that made us feel. Let's call this tool set our **Emotional Tools**. These tools include fear, guilt, peace, shame, and many more.

The third set of tools is generated by a combination of our physical and emotional toolset. Let's call these tools our **Secondary Emotion Tools**. These are the emotions that come after the original emotion. These include hatred, distrust, anxiety, depression, etc.

The fourth set of tools are **Dreaming Tools** that create perceptions of sight. This perception is independent of the sight tool (the eyes), yet provides a form of visual experience to the person dreaming.

## What can we learn with these tools?

I can perceive that there are other perceptions other than my perception. When I talk to other people, they can communicate their perception. I can understand parts of their perceptions because I perceive similar things. For example, we both see a dog and can agree we can both see this dog. Therefore, I can conclude there are other perceptions and I can also conclude there are a very large set of perceptions. Other living things seem to have perceptions. For example, my dog "Bear" seemed to have perceptions. For example, when I came home from work, he could perceive me and show happiness because of his perception of me. This idea can be extended to include all things we call "living." For example, some microscopic animals swim in water and eat and move around objects. They need perceptions to do this. Things we call "non-living" may also have perceptions. Although we have little evidence of this, there is no reason to dismiss this case at this time. Some perceptions are different from different points in space. Physical perceptions seem to be dependent on where I am in space. Other perceptions of mine seem to be independent of my position in space. For example if I am upset or happy, I can experience that perception equally at many different points in space.

We can reach some basic conclusions. There are a large number of perceptions. Many, if not all points in space are a source of possible perceptions. I can perceive some perceptions that are remote from my point in space.

### What is the history of human perception?

A long time ago, humans knew very little. They didn't know how to create fire, didn't know much about the sun, moon, and stars, didn't farm or know about the wheel, but they did have very similar sensing capabilities as we have today. There was a series of understandings that brought us to where we are today without an increase in sensing capability. Simple new concepts brought massive change. Farming brought complex civilization. Wheels brought massive mobility. How about motors, electricity, and one of the newest tools: the computer. With our present senses, our new tools, and new understandings, we will continually increase our local and universal understanding.

As the human race grows and evolves, any one individual will have a smaller understanding of the parts that make up the overall human accomplishment. More and better ways to communicate our understanding to the next generation is extremely important because of the sheer size of information that defines human accomplishments.

### What about the perception of life on Earth?

We know life on this planet had many "faces" over the history of life here.

The local abilities and strategies of life were very different over the last few millions of years, but given our present level of understanding of past life on Earth, all of the brains, senses, and body types are similar for all present and past forms of life. One gets the perception that of the millions of types of life in many different eras, they basically evolved to be very similar.

**Are there other living things in the universe?**

Now some of you may disagree with the conclusions reached here, but don't let it stop you from continuing to read the book. It has very little impact on the future projections.

Let's use an example. Let's say a close friend and someone you trust implicitly showed you a beautiful large grain of sand that was pink with purple stripes. Your friend said he found it on Daytona Beach close to the waterline. You would like to have one for yourself but your friend won't give it to you, so you take the grain of sand to a geologist to verify that it truly is natural and not man-made. The geologist confirms the grain to be natural. Off you go to Daytona Beach and sit yourself down on the sand close to where your friend found the pink with purple striped grain. You fill a large beach bucket with sand and return to your room and dump the contents on the kitchen table. For the next two days you examine each grain of sand and to your chagrin you find no sand that is pink with purple stripes. You get another bucket of sand and repeat the process. After a year of doing this, you feel it is time to draw a conclusion and that is: There is only one pink-with-purple-stripes grain of sand on this Earth.

Even though a lot of work had gone into the attempt to discover this particular grain of sand, most would not agree with this conclusion.

From our present understanding, it would seem odd to think we are unique in a universe that is so vast. We know there is at least one example of life in the universe and that is us. The number of stars that we can confirm with the Hubble telescope is vastly larger than the number of grains of sand on all the world's beaches. Only recently have we discovered over 300 planets that are very similar to ours. Let's say we have looked at one trillionth of the universe. Then it would be reasonable to say there are 300 trillion planets like ours. We aren't seeking something we have never experienced. We are seeking something like us that we know to exist in the universe. So is it reasonable to think we are unique to the tune of one in 300 trillion?

Some of us just don't want to believe there could be other life in the universe for many personal reasons. That is fine, but given our present way of predicting the future with statistics, it is almost assured there are other forms of life out there somewhere.

Once you postulate there are other forms of life, then the next question is how many separate examples of life exist? Given the trillions of places where life could exist, it is reasonable to pick a large number. By the way, we have only checked less than a few dozen planets or moons.

Assuming a large number of places with life in the universe exists, what are the chances we are the most advanced? Well, how long have we been advanced? The Earth has been here 4.5 billion years and simple cell life has been on Earth for 3.8 billion years and human-like creatures with abilities to manipulate their environment have been here about 200,000 years. Just taking the Earth model, it takes about four billion years to get animals like man. But once they evolve, intelligent life, like humans, grow their abilities to manipulate their environment extremely quickly. Any significant human technology advancement happened in the last 250 years. Given that a large part of the universe is older than our solar system, it would be reasonable that some life is more advanced than us.

More advanced forms of life have probably learned about the questions we still have about ourselves and our surroundings.

I am going to conclude that with our present understanding, there is other life in the universe and some of it is more advanced than we are. You may not feel comfortable with this, but it is an important perception to continually evaluate. As I mentioned earlier, life beyond Earth is not the basis for any of the projections that will be made.

Let's now address the present perception of genetic engineering—or as some say, "Man messing with Mother Nature."

Science has brought us to the point where we can transfer genes from one species to another, so that we can change the traits of plants and animals. Assuming that these changes are done with great care, we can develop foods which are more productive, more nutritious, tastier, and better for the environment. It is also possible to proceed carelessly and do damage to the environment and to people's health. The record so far is that the scientists developing transgenic crops have been very responsible and the regulatory agencies have been very cautious. Millions of liters of pesticides have been left in their barrels instead of sprayed on fields and millions of cubic yards of topsoil have stayed on the fields instead of choking streams and rivers. Still many oppose the use of transgenic food. As long as the movement to frustrate genetic-engineered agriculture remains effective, investors will seek other directions, and young scientists will choose to work on less controversial research.

The most successful kind of medicine has always been prevention rather than cure. Genetics is no different and the hope of replacing damaged DNA by gene therapy is still where it has been for the past ten years. Genetic surgery, the ability to snip out pieces of DNA and move them to new places has great potential, but we will not realize that potential until our society agrees to allow it to happen. This is social perception at work and is very important for human development.

Here are two examples of genetic engineering being used today.

1) Hard cheeses are made from whole milk by adding an enzyme called chymosin (rennet), which was formerly extracted from the stomachs of calves. The gene for making chymosin was transferred from cows to yeast. This yeast with the gene for making chymsin is now grown in vats and used to make hard cheeses. Chymosin from yeast is cheaper and purer than chymosin from calves. So today,

almost all hard cheese (over 90%) is made from chymosin produced by genetic engineered yeast.

2) Cotton farmers are plagued by various insect pests, such as the boll budworm, the tobacco budworm, and the pink bollworm. In the southern United States, where most of our cotton is grown, these insects were controlled using chemical insecticides. But there is a natural insecticide which has been used for almost a century by organic farmers, a bacterium called *Bacillus thuringiensis*, or "Bt" for short. The bacterium produces a toxin which is deadly to caterpillars like the three mentioned above, but harmless to almost everything else. Even the legendary boll weevil is not harmed by the Bt toxin. So genetic engineers transferred the gene for Bt toxin from Bacillus thuringiensis to cotton. The cotton plants, which could make Bt toxin, were cross-bred with other varieties in the old- fashioned way. Today, much of the US cotton crop is genetic-engineered for the Bt toxin trait. The use of chemical insecticides in the Cotton Belt has declined dramatically. Since the Bt toxin is inside the plant instead of sprayed onto the plant, the only insects which it can harm are those which eat the plant.

Tests are commercially available for genes predisposed to cystic fibrosis and breast cancer. These tests or "DNA chips" can screen many genes at once and many more will soon be on sale at your local drug store. Medicine will have to deal more and more with those who have diagnosed themselves as at risk.

Okay, so much for perception for now. It boils down to two separate forms of perception we will address. The first is "social" perception and this perception will either allow or stifle the growth of new discoveries. Today social issues in the news are things like cloning, using stem cells, and organic foods.

The second form of perception is an individual's "personal" perceptions. Understanding this perception will be a key for developing out future projections.

Chapter 9

# Connections Between These Fields of Interests Today

Today there is a lot of interaction between our physiology and our perceptions. The primary human effort here is our work in the field of medicine. Tools and processes available today allow heart and other organ transplants, stents, grafting skin, prosthetics (which include robotics), electrocardiograms, electroencephalography (measuring the electrical activity in different parts of the brain), the whole field of psychology, and more.

Of course, there is a lot of interaction between computers, wireless communication, and robotics. Some examples include automation in manufacturing—things like OnStar or SYNC for cars, cell phones, home and mobile PCs, GPS in various applications, the internet, etc.

Our present mathematics and physics seem to work well with computers, wireless communication, and robotics but have trouble explaining how our physiology and perceptions work.

Any time in history, humans have believed or made up stories to believe they knew all about themselves and the world they lived in. About five hundred years ago, we all believed the Earth was flat. At one time, we thought flying machines were impossible. More recently, some scientists believed we couldn't go faster than the speed of sound. Today scientists believe things can't go faster than the speed of light.

Based on our six fields of interest and their interactions during the last 30-40 years, here is a partial list of our new tools and discoveries:

- The Internet
- The Cloud
- Facebook
- Twitter
- Texting
- Sophisticated video games
- Cable and direct TV
- Cell phones
- Microwave ovens
- Solar panels in residential homes
- Desktop computers
- Laptop computers
- Touch screens
- The computer mouse
- The Space Station
- Moon and Mars rovers
- The Hubble telescope
- Smart homes
- New medicines
- Car communication like OnStar and SYNC.
- GPS devices for military uses
- GPS for human use with devices for golfers, runners, car traveling, and much more
- Automatic reservation systems

- Organ transplants in medicine
- International 24-hour banking with up time of 99.999%

We have come a long way in only a few decades. The drive and curiosity of humans will continue to develop new and amazing discoveries and we will talk about some of them in later chapters of this book.

Chapter 10

## Why use a period of 20 years before the first projection?

As we enter the projections part of the book, I would like to start by saying a few words about "What is special about 20 years?" Each projection in the next six chapters will be the state of each field of interest in twenty years.

Twenty years is generally called a "generation." It is called that for a few reasons. First of all, there will be a new set of younger people. They will have been brought up with new tools and perceptions that don't exist today and they will have incorporated them in their everyday lives. Today's younger people, who will raise the next generation, will now be generally the best trained and most effective people in the workforce as either individual contributors or managers. These are the people that are making the new tools and teaching their children perceptions that certainly will have changed in 20 years.

Today's older people are now mostly retired and out of the work force. This typically doesn't happen in less than 20 years and doesn't usually require more than 20 years to happen. Every 20 years there is a complete changing of the guard and there will be a new reality based on that change.

Chapter 11

# Human Physiology in 2032

Let's start with the projection for the understanding of the human brain in 2032.

First, let's create a mental picture of the brain. Think of an event that over 50,000 people are attending—maybe a football game, a car race, or a concert. Think of all the people at the event. Think of all the activities these people are performing—group cheering; collecting tickets; finding their unique seats; serving food and drink; security patrolling; groundskeepers cleaning up; people talking to each other; people going to the bathroom, the medical center, the lost and found, will call; traffic control, parking, and more. As diverse as the activities are, they are all happening at the same time with little chaos. Your brain is like the set of all these people and your body is performing all of the activities at that event.

With that picture, let's be more specific. Using the analogy we just presented, your brain is made up of millions of "little people." Now, these little people don't have to be as sophisticated as a complete human being, but they do accept input or information and, based on the information, they may provide an output or activity. Because these little people are part of you and act human-like, let's call them "me humans" or "memens" for short. At first this may sound silly, but it is a slightly different view of how the brain works and requires two new words to explain the model clearly. There are millions of memens that are collectively called your brain.

For every cell in your body that monitors any of your senses (sight, smell, touch, etc.) there is a wire or nerve that connects the cells to a specific memen in your brain. This type of memen is called a **"Monitoring Memen**." Monitoring memens listen to their cell and when that cell is activated (by heat, touch, smell, etc.) it does a "shout-out" that uniquely tells other memens that it has been activated. That's it!!! "**Shout-out**" is the second and final new term we will define and use. In our previous analogy of a sport event or concert, think of a shout-out as being an announcement over the PA system. With one shout-out every memen hears it. Shout outs don't need to be heard by all the memens.

Here are some examples.

At the event somebody wants to get some food. They go to the food concession and ask for a drink and a hamburger. You just made a shout-out to the person behind the counter. The person behind the counter turns around to the workers preparing the food and drink and shouts out "one beer and one hamburger." Maybe there were ten food preparers listening, but the only ones responding were the beer and hamburger preparers. The person behind the counter shouts out "That will be $8.00." It's a very simple concept and I'm sure you could come up with hundreds of other examples. As simple as that sounds and as easy as it is to understand, there is no good computer function that is an equivalent of a "shout-out."

There are other memens listening to these shout-outs. We will call them **Memory Memens**. Whereas Monitor Memens listen to a specific cell and do a shout-out when that cell is activated, Memory Memens listen to a set of Monitor Memens and do a shout-out when they see a specific pattern they have memorized. Memory Memens listen to Monitoring Memens responsible for a specific sense like smell or touch or sound or sight.

There is a third type of memen called "**Control Memens**." For every cell in your body that controls some function in your body, there is a wire or nerve that connects that cell to a specific memen in your brain. These are cells like muscle cells or cells that control the release of enzymes or other chemicals the body needs to stay alive. They do two things and only two things: They listen for other memens to "shout-out" their name and when they hear their name, they activate the cell they control. That's it!!!

A very important point is that there are no "manager memens." There is no reason for the managerial function as we will soon see.

With this information and *only* this information, you can define human activity and here is how it goes.

Let's start by looking at a human at about Week 5 after conception. There is already a brain and spinal cord. The brain has almost all its monitor, memory, and control memens, but they don't have much to do yet. The control memens for the heart and the heart itself are developed enough to start working. The heart Control Memens start randomly working and the heart starts beating. The heart monitoring memens now start shouting out the heart rate. There are a special set of memory memens that are given their memory as part of our DNA coding. They are programmed to have a heart rate of around 120 beats per minute. The heart memory memens hear the monitor memens and shout-out to the control memens to change the heart rate as needed. That's it!! The heart is being asynchronous and the system continues this process for the rest of the human's life.

The human continues to develop following the DNA coding and many other memens start working to control other bodily functions. These functions are called "internal body functions" because they can operate just fine without any interaction with the outside world.

There is a fourth and final type of memory memen that is needed to start and keep the whole human experience working. What if a baby never tried to attempt anything? It never was interested in looking at anything, it never tried to roll over or vocalize. There needs to be something that motivates a baby to be curious and learn and grow. It can be accomplished in a very simple way with this model. There just needs to be a memory or knowledge memen coded for curiosity. Let's call it the "Hey, check that out memen." Whenever a human experiences something significantly different from what it knows, the "Hey, check that out" memen does its shout-out and a focus is brought on that difference. If it is truly new and different, it may be saved in a memory memen as new intelligence.

Let's now move to just after the human baby is born. A whole new very large set of memens start working. These are the monitor memens that sense the external world along with the control memens that primarily control the muscles that move the body in the external world.

So how does this process work? Two characteristics of human babies are arms and legs moving randomly and anything in the mouth is sucked on. The sucking action is predetermined in a set of memory memens and has been coded into our DNA over thousands of years. If this needed to be a learned action, how could you possibly teach a newborn how to suck in order to get food? The arms and legs moving are based on monitoring memens "shouting out" and controlling memens reacting, but in a random way because the learning process for moving the human body around the external world is yet to be learned.

At about six weeks, human babies will seem to stare at objects. This is a team of memory memens learning to focus as a way to get more information about the external world.

At about eight to twelve weeks, the controlling memens for neck muscles coordinate with the seeing memens to allow following of the object that can now be seen in focus.

One of the next steps is the coordination of eye movement. Babies are looking and can focus on things farther away now. Imagine watching a person pacing back and forth about 50 feet away from you. If you wanted to focus on this person, you certainly would follow the person with eye movement and not neck movement. So babies start moving their eyes to follow a moving object or to look at different things. During this learning process, eyes may wander somewhat independently until the ideal process is learned, which is for eyes to move in unison. Eyes moving in unison provide depth perception which is important to "hunter and gatherer" animals.

This process continues in thousands of learning processes, including grasping, rolling over, crawling, standing, and walking.

### A Quick Review

First, memens start monitoring and controlling internal body functions.

Next memens monitor and control external body functions as the external world-sensing organs develop and send their information to their associated memens. This process develops for a lifetime: learning to swim, ride a bike, play sports, play musical instruments, and more.

There is one more very important function provided by memory memens. This function has nothing to do directly with the physical body. These are the memens that develop us based on our specific environment and environmental experiences.

Here is an example. Let's say a 4-year old child sees a horse for the first time in a fenced pasture maybe 100 feet away. Mom picks up the child, points to the horse, and says "horse." The child says "horse" and a set of memory memens associate the vocalization of "horse" with the visualization of the actual horse. Let's say the horse walks over to the mom and child so they are just a few inches away from the horse. All of a sudden, the child smells a new smell and memory memens remember that smell with the object other memens have remembered as an object called a "horse." Let's say the child touches the horse's nose which is very soft. Memory memens remember the touch and associate it with a horse. Let's say that is the end of that "horse" experience and the child moves away so that it can't experience the physical horse. Now, let's say the child is not able to physically see the horse, but by chance puts the hand it touched the horse with close to his own nose. The smell monitoring memens project the specific smell and the memory memen that remembered that smell as a horse smell shouts out "horse." The memory memens remembering horse touch and horse visually provide that association to the child.

The set of this child's "horse" memory memens is the child's set of understanding of a horse.

So far, this set of memory memens has provided the child with a very passive view of the horse. Let's add more memory memen to this child's horse experience.

Let's say mom brings the child to see the horse the next day. As they approach the pasture, the child sees the horse. The visual horse memen shouts out "horse" to all the other memens to let them know a horse is near. The horse approaches within inches. The smell horse memory memen shouts out. The child reaches out to touch the horse's nose. Unfortunately the horse thinks the child is trying to feed it something and accidently nips the child's hand hard enough to hurt badly! The child pulls his hand away and cries.

A new memory memen is taught that when it "hears" a memen shout-out "horse," it responds by shouting out "danger." A whole new set of memens activate after hearing "danger." Some may activate internal memens to increase heart rate in preparation for some physical activity and some may squirt some extra adrenaline. As the danger activity decreases, the "danger" memory memens quiet down—in this case by mom moving the child away from the horse. As the child matures, it will learn to develop its own "danger" response and different humans will have different "danger" responses.

These memory memens and their associated response when activated make each of us a different person.

If I were to stop here, many people would say there needs to be another thing that addresses knowledge of when things happen or when we just think of things independent of our sensing of the external world and our memory.

Examples are things like:

- It's time to start dinner.
- I just thought of a person I haven't thought of in years.
- Where should we go on vacation this year?
- Do I have a doctor's appointment tomorrow?

- Did I lock the door when I left home?
- Did I see my favorite musician in concert in January or February?

Because of their training, computer programmers may want to create another level of control to control all these memen activities. Computer programmers all use this concept when writing programs. This is because the programmer of programs is external to the program so they must tell the program exactly what it's to do before the program starts. But the human is the programmer and the program all rolled into one. This concept is good for humans but not good for computers. How would you like your computer to change its operations based on how it gets to know you by monitoring all of its senses of you (viewing you on a video camera, listening to you on a microphone, knowing what you type into the computer, knowing how you use the mouse, and where you go on the internet). Maybe your computer would develop an attitude toward you based on your behavior with it. There is a high likelihood you would not want your computer to act like this.

I believe this level of control is not necessary with humans. Humans make up their own programs as they go through life. If you like a person, you like them physically or you like how they are "programmed" by their memens.

Let's look at performing a body function you already know, like walking. Because of some "outside world" sense, you want to walk. Your knowledge memory memens shouts out "walk." The body memory memens hears the shout-out and shouts out to the control memens to perform a walk. During the walking process, thousands of body memory memens are shouting out to thousands of control memens in order to walk. These shout-outs deal with balance, moving the legs in a learned walking fashion, swinging the arms and increasing the heart rate if you are walking fast, turning if a knowledge memen shout-outs to turn, and much more. That's about it.

Let's say you don't know how to skip but your friend wants to teach you to skip. Here's why it works: you watch your friend skip. Using an unused knowledge memory memen, you memorize the vision of skipping and create a new shout-out that tells the body memory memens to perform a skipping motion. Now your body memory memens try to mimic the skipping motion as a modified walking motion. In a short period of time, thousands of your body monitoring memens and body memory memens will have learned the process to skip. Whenever you want to skip in the future, a shout-out to skip from the skip memory memen will activate the body monitoring and body memory memens to perform the act of skipping.

Let's now talk about the knowledge side of the human brain that has nothing to do with activating your body. Examples would be learning a number system, learning forms of mathematics, learning processes like how to build a house or weld metals or play baseball or learning a language, memorizing certain past events so as to know how to react in a present event, how you uniquely walk and talk, what you like or don't like. Basically it's everything about you other than your body and what your body does.

Here's an example of how the knowledge side of your brain works. Let's say someone asks you "What is seven times eight?" You know the number system so you know what seven and eight mean. You know the mathematical concept of multiplying and you have memorized your times tables. This means that you have a knowledge memory memen that can recognize "7 times 8" in voice or print and shouts out "56." Another knowledge memory memen hears 56 and shouts out, "Say 56." The appropriate body memory memen hears the shout-out and activates a team of body memory memens and control memens to vocalize the sound "56."

Does our brain have the capacity to perform all these memen/shout-out functions? Let's see.

Here are some facts about our brain:

1. The number of neurons in the brain is about 100 billion.
2. The human brain produces on average an estimated 70,000 thoughts a day.
3. The energy used by the brain relative to a computer is small. The brain uses about as much energy as a 25 watt bulb.
4. The number of synapses in the cortex is about 150 trillion.
5. The number of fibers in a human optic nerve is about one million.

6. The total number of human taste buds (tongue, palate, cheeks) is about 10,000.
7. The number of receptors on each taste bud is about 100.
8. The number of human olfactory receptor cells is about 10 million.
9. The number of retinal receptor cells is about 5 million cones and 100 million rods.

The memen and shout-out process fits well within this capacity of the human brain.

Let's now move on to a different subject. By 2032 genetic research will have discovered much more about the physical structure of human DNA and how to change it with very predictable results.

Just like vaccines have just about eliminated polio, changing the DNA of the mom and the dad that creates the child could eliminate genetic disorders. Ten thousand different diseases have an inherited component in DNA and, at least in principle, we know the genes involved. By the same token, overall human characteristics can be changed to create "custom" children. Today, if you ask a set of parents, "Do you want a healthy baby?" all parents will say "yes." Yet if the child is developing in the mom and it is determined that the child will not be "healthy," many parents will still want the baby because of the concept of taking a human life.

So when does a human life start? Most would not believe that a human sperm or an unfertilized human egg is a human being, but once they unite, a human life is formed.

So changing the DNA of the mom and dad that will unite to create a human life before they are united is not taking a human life and will start to be accepted in society. Just like any other advancement in medicine, genetic engineering will be a powerful form of medicine without having to determine when human life starts.

This uncomfortable feeling of customizing people will lessen significantly over the next 20 years with the result of humans changing humans for the very first time. Somehow we have come to believe there is a natural process that we shouldn't mess with. We call it "the environment" or "Mother Nature" or "the natural world." A general feeling is that we somehow aren't part of it or are attempting to destroy it. There will be a greater acceptance that humans are a very big part of Mother Nature's plan and we will have the ability to accelerate evolution hundreds of times faster than what has happened over the last four billion years. Maybe universally, "Mother Nature's" primary accomplishment after billions of years is to produce a smart organism (like us) that can take over for her and accomplish things much faster and do things that she could not do. A simple example is that Mother Nature could never create a car without help from a living organism. As always, there will be many risks in venturing into this new area and we need to make sure we don't "screw it up."

Genetic engineering among plants and among animals (not including humans) will be common along with genetically combining plants and animals. Plants will be altered to provide human products that now are produced by animals. Horseshoe crabs are regularly captured and their blood is used to make a substance used to sterilize medical supplies. These crabs would be better off if we could copy their genes into a convenient plant or yeast and make the "blood" we need without bothering them anymore. This will be a common occurrence. Plants will make plastics and petroleum and waxes. Many of our medical drugs are now produced by genetically engineered bacteria or yeasts, but in 20 years they will be produced by genetically engineered plants. There are endangered species whose decline has nothing to do with human exploitation. Two such plants are the American Chestnut and the American Elm. Before the twentieth century, chestnut trees were one of the keystone species of America's eastern forests. They were then decimated by a fungus. These trees will be changed to be immune to these funguses and again enrich our forests.

If we want blue jeans, we will create blue cotton. We will get wool and silk from plants. Fruit trees generally grow in warm climates. They will be altered to also grow in much colder climates. Animal genes and bacteria will be spliced to produce organic thread for surgery.

The human race will begin to see that as a part of the natural growth of "Mother Nature," humans will have an active involvement. Humans changing nature will be seen more as the new active addition to natural activity. Humans will willingly embrace these changes and anxiously look forward to new developments based on genetic engineering.

One form of life, like ourselves, growing to a point where we dominate and extend ourselves beyond natural selection processes may be a universal event that has happened many time before in other parts of our universe.

Chapter 12

# The Computer in 2032

Computer operations today are primarily designed to act as a tool to help us do things like solve business issues, entertain us, inform us, and generally amaze us. These types of computers will be even more powerful in 20 years.

But in 20 years, a new branch of computing will be focused on making computers appear to operate like human brains. This will bring a new type of computer to the world. These computers will learn from its inputs like we do as humans. Chip manufacturers (like Intel and AMD or someone else) will be making computer chips based on memens and shout-outs. Operating systems will also be based on memens and shout-outs. Certain types of robots will use this type of computer.

Here is an example of what might happen when you sit down with your "human" computer in 20 years. (This is to make a point, but computers will only act this way when we want them to.)

You sit down in front of the computer with your coffee.

- Computer: You went to Starbucks this morning. Got your normal Cappuccino (Computer smelled it and learned to associate that smell with a Cappuccino).

- Computer user: Yes, I needed it today.
- Computer: From how you look, you drank a little too much last night. I found a new cure for a hangover. Are you interested?
- Computer user: No! But speak quietly. What's important for me today?
- Computer: (speaking quietly) You need to check your financial portfolio. I made some notes on your financial website. Also, it's your sister Judy's birthday.
- Computer user: Call Judie and sing Happy Birthday to her. If she has time to talk to me, get hold of me on my phone.
- Computer user: Anything else?
- Computer: Yes. Your back left tire is low on the SUV and the spa needs more water. You also need a haircut or are you changing your look? By the way, I think your spouse is upset with you for coming home late and your son asked me to tell you he went on a bike ride and will be back at 11 o'clock today.

The whole field of replicating human senses to interface with these computers will be continually developing. New and more advanced senses that extend well beyond what a human can sense will be developed and used with these new "human-like computers."

New wireless communication will exist between these new computers, creating a human-like computer network without people.

Central computers will be collecting information from these networks of "human-like" computers operating in the real world and will be looking to find if they actually learned things we haven't learned. How fast could we learn more about the universe if these networks of "human-like computers" learned quicker and discovered more than the complete human race?

Chapter 13

# Robotics in 2032

Another area of advancement will be in the area of human-like robots.

We already have tools that emulate our eyes (cameras), ears (microphones), sense of touch (touch screens), sense of balance (gyroscopes), and more. All of these tools will be much more advanced in 2032 and they all will have direct or wireless connections to computers—not to the computers we have today, but the human-like computers based on memens and shout-outs.

They will be "learning" new capabilities like spatial recognition, languages, emotions, and controlling their own growth of knowledge primarily based on the tools we give them to sense the external world. Some of these robots will be stationary and some mobile.

In general, these robots with human-like computer "brains" will be mostly experimental, but with great new possibilities. There will not only be technology issues to address, but also social and political issues to consider.

Robots will be more prevalent in our everyday lives in 2032.

Here are some examples:

- In vending machines;
- In car navigation systems;
- In home energy management systems;
- As caregivers;
- Doing very delicate work that requires human-like decision-making;
- Using nanotechnology to perform invasive procedures without surgery;
- As human companions;
- As butlers;
- In our phones;
- In our internets;
- They will be everywhere.

There will be a growing personal connection with those robots that have a large amount of information about you and can relate with you.

Chapter 14

# Wireless Communication in 2032

I say "How are you?" to you and you say "I'm fine today, thanks."

If you think about it, duplicating this in human machines is quite amazing. But we do it every day with phones, radio, TV, the internet, GPS, radar, sonar, etc. The process is humans interacting with machines that communicate wirelessly to other machines that communicate to humans.

Let's look at the parts:

- Machines talking to other machines wirelessly is well-defined, understood, and works well.
- Human-to-machine and machine-to-human communication is right in the middle of a growth period and will be well-advanced in the next 20 years.
- Compare cell phones over the last 20 years. They were big and clunky. They didn't have computers included. Today cell phones are replacing home wall phones, computers, and game devices. Everybody talks about the newest "App." There are "apps" for everything. Bluetooth technology allows hands-free communication and more.

As for 2032 predictions:

- Our present wireless communication will be more powerful, but basically the same as today.
- Wireless communication will move closer to the human body. Many young kids seem to like their skin poked and pierced, so parts of the wireless communication machine will be worn on or under the skin. It will solve the problem of the "cell phone" devices getting lost or stolen or dropped and broken.

As chips and operating systems for "human-like" computers are developed, different types of data need to be transferred wirelessly. This will include new types of data encoding and new types of error checking in order to be most efficient. This will be a whole new field of technology.

The whole world of devices connected with wires will almost be gone. Many homes will have one or two central stations that provide wireless communication to your home devices that provide:

- TV and radio
- Internet
- Home audio and video entertainment
- Appliance monitoring and control
- Home lighting monitoring and controlling
- Heating/air conditioning monitoring and controlling
- Home security
- Vehicle monitor and control and security
- Pool and spa monitoring and controlling

- Monitoring and controlling of the feeding and watering of inside and outside plants
- Gaming stations
- And much more.

The FCC (Federal Communications Commission) will require major revisions to their regulations to deal with the congested airways.

Chapter 15

# Mathematics & Physics in 2032

Mathematics and physics will have little changes over the next 20 years. They will serve us well with our present understanding.

But the work being done in the area of understanding how the brain works will be very interesting to mathematics and physics, as it could present a new way to view our world and challenge the quality of our present mathematics tools.

Simply put, our present mathematics starts with zero and one and grows to explain our local environment. That is also how computers work starting from two states, one and zero. Not very surprising because that's how we view our local environment.

As we learn more about brain operations, we will start thinking of a "brain mathematics" where you start with a concept of everything and whittle it down with "brain mathematics" until you accomplish your goals.

As we get enough knowledge to emulate the human brain with a computer, mathematicians, physicists and computer program developers will begin to form a mathematics that is a better tool to describe the whole human experience.

Communication between "human-like" computers will be very different than how we communicate today between computers. Describing this new communication may require modifications to our present mathematics tools.

Chapter 16

# Human Perception in 2032

Significant changes will have happened in overall human perceptions.

Local cultural barriers will be dramatically reduced. By local cultural barriers, I mean the barriers among the people you know and with whom you associate. Through the growing use of mass communication methods like Facebook, Twitter, email, and other social networks, the next generations will communicate with each other primarily remotely and will know each other by how they define themselves in text, pictures, videos, and sound bites. This will move perceptions from those created in the physical world to those created by other person for you. Since people generally present themselves in social networks as better than reality, you will experience them that way and feel better about others than you might otherwise.

There will be a vastly larger set of human relationships per individual. Facebook-like tools will make these connections. Many of these relationships will be loosely coupled. By that I mean you may have never physically met, heard, or physically experienced many of the people you would claim you know. Many of these people will provide you with different types of knowledge.

There will be a strong feeling of "family" that extends outside the biological family.

Human profiling and judging of others based on physical characteristics will decline. This will include both positive and negative profiling.

## Privacy and Security

Privacy as a characteristic of humans will have diminished significantly. Can you even think of one situation where privacy is necessary for a person? You can see this in the way the younger generation communicates on Facebook. My grandchildren and I are friends on Facebook and when they are talking to their friends, they speak openly knowing I see their communication. I would have never communicated with my grandparents this openly.

Security will be an ongoing issue. Hackers will still be trying to steal your money electronically and there will be many more options available for the bad guys to exploit. For example, let's say you have an application that allows you to track the location of cell phones. AT&T has such an app to allow parents to know the location of their children. Should the federal or state governments have the right to tap into this network and locate you and your children? Bad guys would know if you are or are not home and/or when you will be back.

## And More

People's façades will decrease based on the proposition that you know best how to present yourself to others either for your or others benefit.

People will like to have a group identity over a personal identity.

There will be an acceptance for a closer coupling between man/machine connectivity. Just look how it has changed in just a few years. From talking in person to a person, to talking on the phone to a person, to talking on a cell phone that vibrates or has a personal sound to alert you to a person wanting to talk. From three-channel black and white TV, to color TV, to video games on TV operated with a controller, to TV/human interaction by moving the human body.

Technology companies will support these changes in perception. For example, search engines are changing the way your brain works. When faced with new difficult questions, people would rather search online for an answer than provide their own reasoning. People will be less likely to remember things they can find online. IQ scores will be increasing. Using sources like physical books will decline and be replaced with things like the internet because of the internet's pervasiveness and unlimited accessibility

More excitement in pursuing universal understanding will exist as we discover more about it.

Most humans will accept the perception of other intelligent life in the universe. Simple forms of life will have been found or evidence of their past existence will have been found within our solar system. Humanity will have increased interest in searching the universe for other life forms and learning universal laws and making them part of everyday life.

The concept of man's role in "mother nature" will change from a passive role to an active role.

The general trend will be that of more tolerance and acceptance of each other. This will not come in the form of mandates or moral codes but because of the availability of mass communication tools and a perceived personal value in openness and transparency.

Chapter 17

## Connections between these fields in 2032

The state of the six fields of interest in 2032 will create the following results.

- Communication tools will merge. You will have the same types of communication tools with similar user interfaces in your home, office, car, or on your person.
- Your human-like computers will know who you are and what you want for information and will provide it without your constant inquiring.
- There will be a change in perception to allow world communities to have humans openly manipulating plant and non-human animal DNA for the benefit of mankind. We will understand all the "junk" in human genes. Junk is generally what we haven't yet classified. Some of this "junk" in the DNA might be capabilities we had maybe 100,000 years ago and we could "turn them on" to create enhancements to our present senses or maybe even additional senses.
- Many examples of machines connecting to human nerve systems and augmenting our present senses will be around. Artificial eyes and ears that see and hear. Because of our understanding of how the brain accepts and sends information, the brain will be working in a very

rudimentary way with machines. For example, hearing aids will accept sounds and convert that sound to information directly on nerves, sending the information to the brain in a way the brain can understand it as sound. Similar activity will exist for sight, smell, and artificial limbs.

- Machines connected to a human nerve will trigger brain activity not associated with our conscious being. Things like breathing, heart rate, and signals to organs to release chemically.

- Gene manipulation of parental genes prior to conception and then the uniting of these modified genes to create a human with some previously defined characteristic will have occurred. All first attempts will be for medical reasons.

- A huge issue with human-like computer networks will be what type of controls are needed and who should provide enforcement of the controls. This will be a worldwide issue and a new regulating body will be formed with members from many countries to create a set of laws. This issue will be a common subject in world news.

- There will be a new computer storage method to store and retrieve information learned by the human-like computers. For example, a human-like computer may over time have learned ways to respond to senses and then provide an action associated with the sensation. Things that could be learned are voice communication, words associated with objects and activities, simple mathematics, creating art, and understanding human emotions.

- Human-like computers can share their information or knowledge with all other human-like computers as they experience the outside world and learn. This is a huge step forward.
- Artificial brain models will be implemented with new "human brain" operating systems running on digital computers. Artificial sensing machines will provide most if not all human senses as inputs to the artificial brain. Also, non-human senses will be developed for these artificial brains. Some artificial brain models will be mobile.
- Physical presence will be mainly for physical activities.
- Much more of our military functions will be done with robots.
- New machines will do a better job of modeling brain activity and operation in order to define brain functions.
- Perceptions will be changing about what makes sense to store in the brain and what to store in local or remote computers.

So we see a pattern developing between human understanding of itself and the development of arguably our greatest tool, the computer. This is not a chance happening, but maybe the way the most advanced forms of life in the universe develop. A huge effort will be on making tools that have human characteristics and developing tools and processes to make the human organism healthier, more capable, and happier.

Excitement will be in the air for all. “Mother nature” gave us the bodies that allow us to make and use tools that “mother nature” can’t make. Until the computer, we developed tools to help us do physical tasks better than we could do with just our body. With today’s computers, we can extend the capabilities of our brain to do things faster and with more accuracy than is possible with just using our brains. The new human-like computers connected to robots will be a tool that is better than our natural brain. They will help us understand things locally and universally that either we couldn’t understand with our natural brain or it would take way more time. The human race has never taken on these challenges before.

Chapter 18

## How will these changes affect us and our society in 2032?

The best way to look at the new changes for 2032 is to look back on a change that has happened to any advanced civilization. At some point there was a move away from many rural communities making up the majority of the country to many urban communities making up the majority of the country. Adults, and to some extent children, have to adapt to many social changes when moving from "the farm" to "the city." Maybe you, or certainly someone you know, has done this with varying degrees of success. In any case, it has its challenges. A new, yet similar, change will happen when our society moves from "the city" to "the world network." Whereas the move from farm to city has many physical changes and many mental/emotional changes for the citizens and the society, the move from "the city" to "the world network" will have almost no physical changes but many mental/emotional changes.

Here are some of those major changes:

- As individuals relate, emote, and interact within larger groups of people, there will be a larger focus on your identity within the group and personal development will be focused on adding value within your groups.

- Personal privacy will change dramatically. Today most people that desire personal privacy want it for emotional reasons. Can you think of any real human need for personal privacy? Today we talk about people "putting up a façade" or being "two-faced" or "not being comfortable in their own skin" or "acting different with different people." Through a stronger group identity, people will become less private and take on characteristics like "what you see is what you get" or "I am what I am."

- Personal security will continue to be important, although it will change in the future. Physical security will continue to be important. Mental security will be more effected by your worldwide "friends and family" than your real family and physically local acquaintances.

- The present concept of the biological family will be less important to you and viewed as just a part of your extended family.

- Humans of all ages will be computer knowledgeable and eager to adopt new computer and communication products.

- A new and large concern by all will be who group information and infrastructure. Is it saved to last indefinitely?

- Less of your personal data (pictures, videos, important documents, etc.) will be on your local computer, but will be on a centralized system. It may be a home system or an internet-based system.

- There will be an increased perception that people are pretty much the same from a physical point of view. This will be based on more non-physical communication than physical communication.
- There will be an acceptance of human life as being a combination of both traditional human activity and artificial activities. Today, computer games bring an artificial interaction that isn't a traditional human interaction. It is a human machine interaction. These interactions have been with us for some time, like the "Pac-Man" game many years ago, but today the games are sophisticated with audio, realistic video, sensual feedback through controller, with much more to come.
- We will interact with human-like computers as if they are real people, not because they look real but because they will "know" you and interact with you as a real person would. You could be mentally distraught if you were to "lose" one of these human-like computer friends.
- As our society embraces more new changes from 2032 to 2062, many of humans' current bad personal and group behaviors will be reduced and eliminated as shown in the next few chapters. All this goodness will add new challenges that will dwarf our present human challenges in the area of defining human morality.

The differences in human society between 2012 and 2032 will be many, but the changes will be logical extensions of what we knew before. We can see the projections about us in 2032 as being quite possible. The projections for 2062 will be much different as some very basic human understanding will change. I believe it to be a major step forward in human evolution. We will all be better off because of these changes.

Chapter 19

# Human Physiology in 2062

It is now about time to present the most important advancement in the history of mankind.

As we get closer to 2062 many new things will be quite common. There will be human-like computers with computer chips based on memens and shout-outs. There will be human-like computer operating systems. We will have spent 50 years learning about how the brain works with our body and senses.

It is important to understand we will be well beyond the stages of emulating human-like computers and networks using our existing technology. These human-like computer robot networks are completely different types of computers and networks.

The human body implements the brain/body/senses functions and processes in organic material. The human-like computers implement the brain/body/senses in inorganic materials. There is the likelihood that we have learned how to make pieces of the human-like computer with organic materials. This would be another major accomplishment yet would not change our projections.

We have verified our work by creating the models in our computers and they act very much like humans. We have created many artificial sensors for ourselves and the human-like computers. These sensors smell more acutely, hear and see more acutely and at different frequencies not detectable with normal human eyes and ears. This area of our understanding of our physiology could now be considered mature.

There is one more set of things we are about to discover and it would have been impossible to discover if we hadn't spent the 50 years of discovery in this field.

This discovery and the ramifications of this discovery are really the reason for this book. We will need to provide the projections in the other five areas of interest before further addressing man's most advanced accomplishment ever.

Chapter 20

# The Computer in 2062

Our present types of computers will be doing what they do today but they will be faster, have larger memories, and better user interfaces. If you need to know the right flight path to Mars, you need a computer like we have today. If you want your grocer to add up the cost of your groceries without just guessing, you need a computer like we have today. Our interest in this book is primarily focused on the new human-like computers used inside robots that are wirelessly connected in large worldwide networks.

Human-like computer robot networks will be sensing our world with human-like senses. They also will be sensing the world with sensing devices that far exceed our human senses. Examples would be sensing the whole frequency range from audio to x-rays and beyond, sensing and learning about very small things with electron microscopes and very large things with telescopes.

They will be constantly learning based on new information they will experience through their human-provided sensing devices. They will be networked so they may share the learning of other human-like computers experiencing their "world" in totally different places. Unlike humans, human-like computers don't forget.

Human-like computers can basically "live" forever. The knowledge of all human-like computers will be "backed up" somewhere on the network and if it breaks, gets destroyed, or just gets old, humans can replace it with a new model and download to the new model the old model's knowledge and continue on indefinitely.

By the way, all human-like computers have the exact same potential to be as knowledgeable and therefore as smart as any other. If two human-like computers shared each other's knowledge, each would be as smart as the other.

There would be present day computers as part of the human-like computer networks so the human-like computers could search them for information or do complex calculations on them.

If you haven't already noticed, there is one giant difference between groups of humans working together and groups of human-like computers working together. One reason for humans to work in groups is that different people have different knowledge sets. By sharing all the individual knowledge sets, the group has more knowledge than any individual to make a decision or take an action. With human-like computer networks, any individual human-like computer gaining some knowledge will share it with every other human-like computer on the network. Therefore, every human-like computer in the network will have the combined knowledge of the whole group. This is HUGE. Let's give an analogy:

Let's say that a group or company makes lots of different types of bicycles. There is the President, CEO and Board of Directors. There is corporate management and group management in charge of each bicycle group (kids' bikes, mountain bikes, racing bikes, tandems, etc.). There are marketing, R&D, finance, administrative, sales, and support people at all levels. There are some people with advanced degrees in all different business-related fields.

What if you could take every bit of education in the group, every experience and knowledge of everyone in the group and put that huge set of knowledge in every person in the company? People would still have to do their job (or potentially anybody else's job) but business decision making at any level would be so much better and faster. That would be the power of human-like computer networks.

Let's talk a little bit about the language used in human-like computers in a network. It would be ideal to communicate in the same language. Let's look at the issues with human-like computers in different countries of the world learning with different written and spoken languages and then sharing each other's knowledge.

Some things would be unaffected. Let's say, for example, one human-like computer learned the process of tying shoes or saddling a horse. It could share the knowledge in memens and shout-outs without the use of language. You certainly could learn to saddle a horse without ever knowing the word for "saddle" or even "horse."

Today, to communicate between human-like computers or humans that you saddled a horse would require a common communication language.

There is one more case that requires human-like computers to know a common communication language. It's when they want to invent or understand how tools work. Tools are anything we can make to change how we interact with the outside world. Things like the wheel, the bow and arrow, the wagon, the house, the car, the electric system, or the space shuttle. Tools require some knowledge of the science of physics and the language of physics is mathematics. For humans or a human-like computer to use mathematics, at a minimum they must have a number system and be able to count.

Let's show an example:

- I have a blank piece of 8x11 sheet of paper.
- I have stuck on the paper 43 little colored dots. Some are red and some are blue.
- I show you the paper and ask you to tell me how many dots are on the paper.

If you hadn't learned to count, this task would obviously be impossible. If you have learned to count, you would count the dots in a particular human language like English, French, Spanish, etc. If you lived in France your whole life and only spoke French, then you would count in French. Now, if you came to America and recently learned English and someone asked you to count the dots, you would count in French and then translate the answer to English to give the answer. This would be a complication that wouldn't be desirable in human-like computers. It would therefore be best if all human-like computers communicated and used mathematics with a common language.

At some point human-like computer networks would have discovered things that humans hadn't yet discovered themselves. The power here to help humans accelerate their discoveries is almost "mind boggling" and will be very exciting.

A whole new area that involves sharing and storage of human-like computer knowledge will be a worldwide effort. It will not just be an issue of implementation with computer science, but a social and political issue as well.

For example, what standard language will be used for these human-like computers? English, Chinese, Spanish, German, Arabic, or another? There are different computer standards in Europe than in the United States. Do we create a new totally different language for human-like computers? Do humans try to standardize a language among themselves so there will be no need for the human-like robots to translate between different groups of humans? If a human-like robot in a network discovers something new, who owns that "intellectual property"?

After thinking about this, you may conclude that it will require a new, better, and more cooperative social order. This is now a new starting point for human advancement. Today, it may sound scary, dangerous, out of control, and too risky, but over the next 50 years as it evolves, it will become very exciting and promising to the next generations.

Chapter 21

## Robotics in 2062

Robots of all kinds will be a common part of our life in 2062.

There will be today's type of robots in manufacturing, the military, security, and transportation. They will be in service companies like insurance, real estate, utilities, and many others.

They will also be taking over other responsibilities in areas that we thought were reserved for just humans. Things like companionship, cleaning homes, mowing lawns, cooking, teaching, child care, chaperoning, shopping, being butlers or maids, acting as protectors, and so much more.

The human-like computer robots will be helping us discover and grow our understanding in the area of science, the area of human perceptions, and social behaviors.

Robots in medicine will be everywhere. Robots will be very small and be able to invade your body to perform various medical functions like destroying cancer cells, removing plaque from the circulatory system, killing parasites, fixing ulcers, and more. Robots will colonize heavenly bodies like the moon and Mars. Humans will have to deal with who owns these moons or planets. What about mineral rights? What about moving water and atmospheres to those bodies from resources on Earth? Will robots have a nationality? Will they have an ethnicity?

The definition of a robot is quite unclear and has changed in time and will probably change again in the future. For clarity, let's look at an example using lawnmowers:

- If I am pushing a manual mower to cut the grass, is the combination of me and the lawnmower a robot?
- If I am riding a lawnmower with an engine to cut the grass, is the combination a robot?
- If I am sitting on the porch drinking an ice tea and my son is riding a lawnmower to cut the grass, from my point of view is my son on the riding lawnmower a robot?
- If a computer is operating the lawnmower to cut the grass, is the combination a robot?

Where do you draw the line regarding what is robotic? As human-like computer robots begin using tools, do we need to make a distinction between humans using tools and human-like robots using tools? The reason for this exercise is to present the chronological growth of humans' abilities using their body, their tools, and their robots. Here is that chronological order from earliest to 2062:

1. Humans using their unique body as a tool. Things like standing upright to free our hands that have an opposable thumb which allows us to manipulate our world like no other animal. This enhances our ability to manipulate the physical world that we experience through our senses for our benefit.

2. Humans building tools from materials found in our world. Things like clubs, bow and arrows, simple boats, fire making tools, and more. This enhances our ability to manipulate the physical world that we experience through our senses for our benefit.

3. Humans making materials like bronze and iron that allow us to create many more new tools. This enhances our ability to manipulate the physical world that we experience through our senses for our benefit.

4. Humans making and using material to create electricity and electronics along with the tools that use electricity and electronics. Things like motors, trains, planes, and automobiles, radios, TV, radar, and so much more. This enhances our ability to manipulate the physical world part of which we cannot experience through our senses. Humans cannot sense electricity unless they touch a live electric conductor. Humans cannot experience communications going across a telephone cable or TV signals in the air. This is quite phenomenal!! By playing around with tools, we can actually experience parts of the physical world that our body has no way of sensing. This is a major step that is very different from the first three steps.

5. Humans making today's computers. This is the first tool that works both in our physical world and our mental world. Today's computers provide us solutions to questions that humans alone could never answer. This is a major step that is very different from the first four steps. This tool allows us to grow our understanding beyond the

capability of our collective brains. As powerful as computers are today, they are highly controlled by humans. Today's computers don't answer questions we don't program them to answer. In other words, they don't come up with original ideas. At this point in time, all original ideas and original discoveries are done by humans.

6. Humans making human-like computer robots. These tools can now learn like humans and discover like humans. They can also learn and discover in areas where humans can't sense with their bodies. This would be using electric, electronic, and computerized tools. This will open a vast new world of discovery and learning for humans with a tool that is totally independent of us physically and mentally. This tool dwarfs the capability of any previous tool. There is a possibility for these tools to discover universal "laws" that we so sadly lack today.

Chapter 22

# Wireless Communication in 2062

The concept of wireless communication will be mature from a speed and connectivity perspective. The real changes will be in how to move human-like computer information from one place to another. Moving memens and shout-outs will be much different than how we presently move words, pictures, videos, and other binary data. We will have to learn how to bundle human-like knowledge from one human-like computer and send it to another where it will be unbundled and turned into knowledge again.

There is the subject of data storage of human-like computer knowledge. It certainly needs to be archived, protected, and redistributed. Where will it be and what are the requirements to share? Who will own, control, and give access to the archived knowledge? This is just another area that may help humans to work together for a common cause. The cost to a group of humans not being involved with these new discoveries will certainly affect the value and abilities of that group.

Chapter 23

# Mathematics & Physics in 2062

There are always side effects and new unsolicited discoveries when you enter a brand new field. Human-like computers will be one of those brand new fields.

Our present mathematics is very structured and repeatable. Mathematicians would all agree that present computers work as well as they do based on present mathematical understanding. Mathematicians would also agree that computers don't act like humans. So when we build and advance the study of human-like computers, do we need to add or change our mathematics accordingly? Humans have created a massive amount of order in our world, including computers and mathematics, and we did it with a mathematical tool provided by our brain. We have continually changed our concepts in mathematics to represent and document the world we perceive and how it works.

25,000 years ago, we were quantifying time. 5,000 years ago we invented counting. 1,000 years ago, we invented the first computer called the abacus. In 2062 we will have unlocked the mathematics used by the human brain. Of course, nobody knows today what that mathematics will be, but humans will be more understanding and interested in universal laws and universal tools. Things that are true throughout the universe and always have been and always will be. Maybe a simple answer is to simply remove our concept of time from our present mathematics. Anything universal should be timeless. Our new human-like computer robots may help us with this discovery.

Chapter 24

## Human Perception in 2062

Our tolerance and acceptance of each other will continue to grow, but now it will be fueled by necessity. Everyone will see the value of standardized communication languages, standardized human-like computers, standardized mathematical tools, and shared information in a universal form because of human-like robots. Cooperation within a group when each member of the group has a compelling personal reason to cooperate will drive this standardization to be implemented and used by all members of the group. There will be a general sense of major accomplishment for all humans and an exciting pioneering feeling as we continue to learn with our new tools.

Many new challenges will be emotionally charged. As we tend toward international cooperation, various feelings will emerge as we address issues like these:

- How to address a totally new set of feelings and perceptions when one of our tools, human-like computers installed in robots are more mentally capable than us.
- How do we deal with a tool that we, by design, don't control and it decides to "behave badly?"
- How do we define this bad behavior and keep it from spreading among the members of a network?

- How would we react if a human-like computer network decided to not communicate with humans anymore?
- How will we deal with the social structure developed by human-like computer networks?

Of course, there is the area of genetics and how new discoveries and new uses of genetics will affect our perceptions.

- How will we feel if DNA is found to be a universal coding for life?
- How will we feel as we modify plants and animals for our own reasons? What if we could have "plants/animals" create a gas/oil product in high volume while removing $CO_2$ from the atmosphere and adding oxygen?
- How will we feel as we eliminate hereditary diseases?
- How will we feel as we eliminate hereditary unwanted birth defects?
- How will we feel as we change human members of a group to be physically and mentally capable for specific tasks?

It will be a very different world which will require the highest degree of change in human perception we have ever experienced. The human-like computer robots will be far better teachers than we have been to ourselves and we will accept that it will be beneficial to our continued growth.

Chapter 25

# Connections Between These Fields of Interest in 2062

The convergent of the six fields of interest in 2062 will provide one of the greatest steps forward in human existence.

Humans will not just be using their given senses to live in their local environments.

Humans will not just be understanding the environments where they live and creating tools to advance how we can manipulate that environment to their liking.

Humans will have unraveled one of the most complex things in the universe. They will have learned how their brain works and how to use it as a tool to enter a new level of understanding on Earth and beyond. We will now have a tool we created that augments our mental capacity and allows us to learn more at a faster pace than ever before. In some cases, our human-like computer robots will make discoveries that the human brain would never be capable of discovering.

We also will be using our understanding of our DNA in many different ways: two examples are preventive medicine and health management. We will be creating new plants, animals, and combinations of plants and animals to help us produce sources of food, medicine, or new materials to be used throughout our infrastructure.

Humans' perceptions over this vastly changing half a century will change in ways to foster and grow a worldwide desire to cooperate with one another in almost all aspects. By no means does this mean we will be less diverse, but exactly the opposite. We will be a much more diverse set of people, but with true respect for each other's contribution and position. Issues of crime will be greatly reduced because of the demeanor of people, not because of policing efforts. There will be few reasons to be warring and world-related issues will be addressed mentally instead of physically. Our emotional relationships will consist primarily of positive types of emotions. If you exhibit negative emotions toward others, you will not be supported by others.

It seems amazing that these new human tools can change the very nature of humans so that we are better off as a group.

There is one more extremely important discovery that has brought us to a key turning point for mankind. A totally new area of human development will be available for us to use if we feel it is right.

The final chapter will present this discovery.

Chapter 26

# The Big Dilemma

After all these projections become a reality, one thing is clear and that is we will be a better human society. People will be more tolerant of each other, people will view knowledge as a commodity, most people will be contributing to society from home as long as they aren't making physical things that human-like computer robots cannot make, new world order will be defined in a way that focuses on cooperation from all humans with peaceful processes to address differences, and much more. Humans will have to determine what they will keep as their functions in a world where human-like computer robots can perform most human functions better than humans.

This will not be a boring environment for people. Most assuredly, it will be the opposite. Humans will have much more time to indulge themselves in the emotional endeavors they enjoy. With a society where everyone is highly knowledgeable in their field, results happen in less time including necessities of life, like obtaining food and shelter. Formal working time may be only a few days per week.

Surely philosophical questions will have to be evaluated. What is it we should do? Our human-like computer robot networks will be presenting new information for us to digest at an alarming rate. How should we react to this new world?

Given all of this, we still haven't talked about the most pressing and possibly troubling issue that we will need to address.

All of the discoveries in the six fields of interest we have discussed will allow us to make a major discovery that will present humanity with a major dilemma. Here it is.

Chronologically it happened like this:

- We have learned how to make tools that emulate our senses and how to connect them to our body. Things like hearing aids, artificial eyes, artificial limbs, artificial vocal cords, and other senses beyond our human sensing capability will be connected either wirelessly or wired to our nervous system and will be recognized by our brain just like our other human senses.
- We learned how our brain works and duplicated how our brain works in a tool.
- We have learned the physical process of shout-outs. We have learned the language of shout-outs. We have learned the way memens recognize shout-outs. We have learned how memens store a particular incident of a sense. This field of study is what grew and developed our human-like computer robots.
- We have developed the wireless network that can transfer memen information from one human-like computer to another. This will be a new network with a new set of

protocols that will be very different from what we use today.

- We have developed the language for human-like computers that mimics the language the human brain uses. Human-like computer robots com-municate with each other and share information using this language.
- We have achieved the human-like computer "knowledge" within the network in a way that other human-like computers can access it and use it as they see fit.

Along with this will be the parallel development of human "telephones." As our understanding grows in the development of human-like computers over the next 50 years, these human telephones will develop so humans can communicate with the human-like computers. There will be a common language we learned from understanding brain communication that we use on the human telephones. Humans will also have access to all human-like computer knowledge using human telephones.

At this point, only one part of the human/human-like computer robot network is missing. We will have developed the capability to implement this missing part, but should we? That is the big dilemma.

Should we connect humans directly to the human-like computer robot network? We will have learned how to connect to our human memens wirelessly with the human telephone in such a way as to transfer memen and shout-out information in both directions between humans and the human-like computer network. Our human telephones will transfer a new sense to the human brain that we will learn to use. It will be like a computer screen, mouse, and keyboard that is visualized only in the mind and can be operated by the human brain.

Humans will now be able to be a fully integrated member of the human-like computer robot network.

Okay, so why would that be such a big deal? Here are only a few of the reasons:

- Let's talk about human learning. If you wanted to learn your multiplication tables, you simply would download that knowledge to your memory memens from the network archive. Maybe it would take five minutes. Maybe you would like to earn your doctorate degree in physics. That download may take an hour.
- What about learning to play the piano by downloading knowledge that was uploaded by a concert pianist? The issue here is a physical issue. Maybe your fingers aren't long enough or your reactions are too slow. So learning complex physical tasks would be difficult. Of course, if humans decided to clone certain people to be similar physically, then complex physical skills could be learned by downloads.
- What about learning complex processes that don't require complex physical skills? Things like tying your shoe, saddling a horse, learning to program a computer, changing a tire on your car, learning to be a gourmet chef. This type of process learning could easily be transferred by a download.

- What about uploading your complete knowledge and emotional information to the network for archive? Could you download that same information to another human body or human-like computer robot? Would that human or human-like machine be you just in a different physical body or machine? This would mean a person could live forever past the physical demise of a physical body? Does this mean we would have to redefine the meaning of death? What about the meaning of time if one can exist in some physical body forever? What would be a lifetime?
- What about taking a vacation to Hawaii by uploading and downloading your mental information to a human-like computer robot or other person already in Hawaii? Maybe you would like to vacation in Hawaii as a beautiful young woman or handsome young man.
- What if there is other life in the universe that has listened to our wireless communications and realized we haven't yet discovered how our brain works just by listening to our communication protocol? They would assume we are still underdeveloped warring animals they wouldn't want to contact. Once they notice our wireless communication protocol is based on an understanding of an advanced animal brain, we may get a phone call or visit from beyond. Maybe we could upload one of us to their network to be downloaded to one of them and vice versa.
- Could we add dolphins to the network?
- What if you as a male have some downloaded knowledge that is specific to a female human body. How

would that work? This is a key point that could be realized in many different scenarios. It also is a totally new concept to address. First of all, is it imperative to have the memory of the physical activity that generated the knowledge? The answer is "No." We have many memories that we can't recall exactly how we got the memory. But if we recalled a memory that would be unique to a sex that wasn't your sex, wouldn't that cause a giant form of confusion for you? Yes, it would, but it could be a learned trait to ignore the physical connection part of a memory as this could be a new thing we as humans would have to deal with moving forward.

- Our real and emotional attachment to our physical body will be very different. Our mind body experience will be a short-term set of experiences in the continuum of the growth and evolution of our mental capability that could be saved and moved to another body.
- Would you want to repeat a normal human life every 80 years or would you like to be in your 20s forever?
- What will be the meaning of human life?

Humans will start a new era of our growth and development. We will move from a primarily physical world to a primarily mental world. Of course, the physical world will still be very important. For example, humans must continue to reproduce themselves. It will be a long time from 2062 before human-like computer robot networks will clone and raise humans.

It will be a wonderful journey to a better place for all our future generations. Wait a minute, soon there will only be one generation.

www.ingramcontent.com/pod-product-compliance
Lightning Source LLC
Chambersburg PA
CBHW060623210326
41520CB00010B/1450

* 9 7 8 0 9 8 8 6 5 4 4 3 3 *